Vitor Rocha dos Santos
João Bosco Faria

Use of sugar in cachaças obtained traditionally and by redistillation

Vitor Rocha dos Santos
João Bosco Faria

Use of sugar in cachaças obtained traditionally and by redistillation

Effect on the sensory quality of the drink

ScienciaScripts

Cover image: www.ingimage.com

This book is a translation from the original published under ISBN 978-3-330-99820-9.

Publisher:
Sciencia Scripts
is a trademark of
Dodo Books Indian Ocean Ltd. and OmniScriptum S.R.L publishing group

120 High Road, East Finchley, London, N2 9ED, United Kingdom
Str. Armeneasca 28/1, office 1, Chisinau MD-2012, Republic of Moldova, Europe
Managing Directors: Ieva Konstantinova, Victoria Ursu
info@omniscriptum.com

Printed at: see last page
ISBN: 978-620-8-61902-2

DEDICATE:

I dedicate this work to my parents Rosely and Lairton, for their support, encouragement, love and for the teachings that formed the foundations of my story.
To my grandparents:
Natal Joaquim dos Santos (in memoriam) Satyro Rocha da Silva (in memoriam).

ACKNOWLEDGEMENTS

To the Faculty of Pharmaceutical Sciences/Unesp, for the opportunity to carry out the work and the knowledge acquired.

The journal Alimentos e Nutrição, where part of the work was published and which allowed it to be used for this book.

To Capes for the financial support.

MAPA (Ministry of Agriculture, Livestock and Supply) for the physical and chemical analyses.

SUMMARY

Sugarcane spirit is the second most consumed alcoholic drink in Brazil, with production estimated at around 1.2 billion litres per year. In 2009, 10.83 million litres were exported, generating revenues of US$15.58 million. Among the main destination markets are Germany, the United States, Portugal and France (ALVES, 2009).

Faced with the new demands of the market, producers are starting to worry about adding more value to the drink, investing in new technologies and staff training, in quality control and offering more sophisticated packaging, with the aim of placing cachaça in the country's best consumption points, alongside other distilled drinks that are already well-established in the country and the world.

The addition of sugar has been used as a way of standardising the taste of some distilled beverages in order to reduce the taste of ethanol and any sensory defects (CARDELLO and FARIA, 2000).

The aim of this study was to assess the effect of added sugar on the sensory quality of cachaça by comparing traditionally produced and redistilled samples containing different concentrations of sugar.

Keywords: Aguardente, sensory analysis, cachaça, redistillation.

SUMMARY

INTRODUCTION AND JUSTIFICATION

Sugarcane spirit is the second most consumed alcoholic drink in Brazil, with production estimated at around 1.2 billion litres/year. When the distilled spirits market is considered exclusively, it ranks first in the country (87% of the market). In 2009, 10.83 million litres were exported, generating revenues of US$15.58 million. Among the main destination markets are Germany, the United States, Portugal and France (ALVES, 2009).

However, the volume of cachaça exports still represents less than 1% of the total produced (ALVES, 2009).

There are currently more than 40,000 producers (4,000 brands). Micro-enterprises account for 99% of all producers. Their agricultural activities include the production of corn, beans, coffee and milk, among others, and the production of Cachaça. The Cachaça sector is responsible for generating more than 600,000 direct and indirect jobs (ALVES, 2009).

Brandy is highly prized because it has a characteristic aroma and flavour, which is often modified by storing it in wooden containers, which transfer compounds in its structure to the drink, improving its sensory quality. This modification depends on the storage time and storage conditions, determined by the characteristics of the barrel and the environmental conditions of the cellar. Cachaça production in Brazil is an important industrial sector and a source of employment (CARDELLO 1998).

The sugar cane brandy industry, when well implemented and technically exploited, represents one of the best rural ventures in terms of profit, reaching both domestic and foreign markets (SEBRAE, 2008).

Abroad, cachaça has been making a name for itself in some European countries, such as Germany, Italy and France; North America, such as the United States; and Asia, such as Japan, where it is consumed in cocktail form, especially the caipirinha, which has become a "trademark" beyond our borders (NOGUEIRA, 2005).

However, the volume of cachaça exports in 2006, which totalled 11.3 million litres, is still derisory when compared to the total produced, representing less than 1% of the total produced (SEBRAE, 2008).

Exports are low mainly due to the fact that Brazilian brandy is not of the highest quality, mainly due to the technical conditions of this agro-industry, which leave a lot to be desired, apart from a few rare exceptions. Evidently, the reason for this lies in the lack of basic knowledge on the part of those responsible and the technology that would allow them to obtain a product of superior quality, combined with a better industrial yield.

Faced with new market demands, producers are starting to worry about adding more value to the drink, investing in new technologies and staff training, quality control and more sophisticated packaging, with the aim of placing cachaça in the country's best drinking spots alongside other distilled drinks that are already well-established in the country and the world.

The addition of sugar has been used as a way of standardising the taste of some distilled beverages in order to reduce the taste of ethanol and any sensory defects (CARDELLO and FARIA, 2000).

In the case of cachaça, Brazilian legislation allows the addition of up to 6g of sugar per litre, as a normal standardisation procedure for this drink. However, when the term "sweetened" is stated on the label, the addition of sugar can reach up to 30g/L (BRASIL, 2005). In this case, it is worth questioning whether this amount does not go beyond standardisation and become a way of masking the flavour of poorly processed cachaças.

This practice is also related to another technological problem that has worried many producers, namely the subsequent appearance of flakes in spirits that have already been bottled and placed on the market. This defect, which compromises the appearance of many alcoholic drinks, can have various origins (SIELBERG, 1999), but in the case of Brazilian cachaça, it is mainly related to the practice of adding sugar to the distillate (ISIQUE, 1999).

OBJECTIVES

General:

To evaluate the effect of added sugar on the sensory quality of cachaça, comparing samples obtained traditionally and by redistillation containing different concentrations of sugar.

Samples were compared from a sensory and compositional point of view.

Specific:

- To check the effect of added sugar on the acceptability of samples of traditional and redistilled cachaça.
- Compare the profiles of volatiles and any contaminants present in the traditional and redistilled samples.
- Determine a minimum standard quantity of sugar to be added to cachaça in order to improve its sensory profile.
- Compare the effect of redistillation on some physical and chemical parameters.

CHAPTER 1

LITERATURE REVIEW

HISTORY OF CACHAÇA

Sugarcane brandy appeared in the early days of our colonisation, as soon as sugarcane plantations were planted and sugar mills began to grind sugarcane, and it has played an important role in our economy ever since (BIZELLI et al., 2000).

The first Brazilian brandies were obtained from sugar molasses, the residue of the mills set up here by the Portuguese colonisers. The first of these was built in Santos, SP, by Martim Afonso de Souza, grantee of the Captaincy of São Vicente, in 1534 (LIMA, 1999). It is not known, however, when the Portuguese, who knew the technique of producing bagaceira, improvised the first stills to distil the fermented molasses or where, on the vast coast of Brazil, the first Brazilian cachaça was obtained.

According to Lima (1983), until the end of the Second World War, the cachaça industry was essentially rural, involving a large number of rudimentary, technically backward factories with small production volumes. Generally, the factory owner himself, with the help of his family, planted the sugar cane and commercialised the brandy he produced. Few producers bottled their product and there were hardly any exclusive bottlers.

The post-war increase in population and consumption, together with the habit of drinking cachaça, already incorporated into the Brazilian way of life, led to an increase in sugar cane plantations and the production capacity of distilleries. The small stills began to give way to large installations equipped with continuous distillation columns and large bottling plants emerged, with their own commercial brands, which are now spread all over the country (LIMA, 1983).

CACHAÇA PRODUCTION

CANE BRANDY

Term used to generically describe Brazilian distillates obtained from sugar cane (BRASIL,

2005).

SUGAR CANE

Sugarcane (Saccharum officinarum) is one of the most widely cultivated grasses in the world's tropical and subtropical regions, due to the enormous socio-economic contribution its exploitation makes, as a result of the plant's ability to synthesise and store a significant amount of sucrose in its reserve tissues. (NOGUEIRA & FILHO, 2005).

According to Novaes (1995), there are no sugar cane varieties that are specific to the production of brandy, as the small differences in their physical and chemical composition would probably have no influence on the final quality of the drink. Sugarcane varieties must be adapted to the climatic conditions of the region where the industrial unit is located, with the main aim of having a high sugar yield per area.

In the production of brandy, the raw material of interest is sugar cane stalks, which consist of two phases: the solid and the liquid. The solid phase, a complex made up of cellulose, lignin and pentosans, is generally referred to as fibre, and the liquid phase, or juice, is an aqueous solution containing a wide variety of organic substances, approximately 90% of which is sucrose (NOGUEIRA & FILHO, 2005).

PROCESS STEPS

MILLING

Extraction is normally carried out by directly crushing the cane in mills that must provide maximum sugar extraction. The future of the company depends to a large extent on the right choice of mill, since the factory's yield will depend on its extraction efficiency (LIMA, 1983) and its good performance will ensure the uninterrupted operation of the industrial process during the harvest (FARIA, 1995).

Sugarcane juice, as obtained from the mills, can be considered a natural must, since it is a liquid that can be fermented and, by definition, is called must. However, the industrial fermentation involved in producing brandy, as well as guaranteeing a good quality product, must also provide economically viable conditions for obtaining the product, which will require the must to fulfil certain conditions that the juice that comes straight from the mills does not. Thus, the preparation of the must consists of a series of measures to be taken so that the fermentation medium becomes favourable to the alcoholic yeast whose action is desired and is contrary to the development of undesirable microorganisms (LIMA, 1983; FARIA, 1995).

Less initial contamination of the must can be achieved by adopting hygienic practices at all points of contact with the juice (conveyors, mills, vats, pumps, pipes) and by decanting the juice after milling, which favours the separation of impurities such as sand and earth.

Conditions favourable to the development of yeast and its maximum fermentative activity can be achieved by diluting the broth, because when it is too rich in sugars, the production of ethanol at certain concentrations begins to inhibit the action of the yeast; as well as controlling the temperature at around 30°C and the optimum pH, which is between 4.5 and 5.0 (FARIA et al, 2003).

FERMENTATION

The fermentation of sugar cane juice is mainly alcoholic and begins the moment the must comes into contact with the yeasts present in the inoculum, the solution that contains the alcoholic yeast in suspension and represents around 10-20% of the total volume to be fermented (FARIA, 2000).

Generally, the fermentation process lasts an average of 24 hours, with recycling of the inoculum, i.e. after this time, the yeasts will settle at the bottom of the fermentation vat, when a new must will be added, the temperature of which should be around 30 °C (CANTÃO, 2006).

DISTILLATION

After fermentation of the sugar cane juice, the must is called wine and has a variable composition of gaseous, solid and liquid substances. Carbon dioxide, which remains dissolved in small proportions in the wine, is its main gaseous component. The solids are basically made up of yeast cells and bacteria, mineral salts, sugars that have not fermented and suspended mechanical impurities. Finally, making up the liquid phase of the wine are water and ethanol, its most important components from a quantitative point of view, and in smaller quantities, a series of other so-called secondary compounds, which are mainly responsible for the aroma (NÓBREGA, 2003) and flavour typical of sugar cane spirits (YOKOYA, 1995; FARIA et al, 2003).

The wine is distilled in stills or columns of various types and sizes, obtaining the simple alcoholic distillate of sugar cane, which can then be diluted, or directly into cachaça (FARIA, 2000).

The distillation of wine will therefore promote the separation and concentration of the compounds present, which through volatilisation or vapour drag, followed by condensation, will make up the distillate, the composition of which in turn will depend on numerous factors, including the type of distiller used, the way in which the distillation was conducted and the distillation time (LIMA, 1964).

Distillation to obtain cachaça must be fractionated into three distinct parts to remove undesirable compounds:

Head: this is the first distilled fraction and contains most of the methanol, so it must be separated from the final product.

Heart: this is the second fraction, the brandy as such.

Tail: this is the last fraction and, like the head, contains undesirable compounds (less volatile products) and must be separated from the final product (FARIA et al.,

2003).

BIDESTILATION

Bi-distillation of sugarcane spirits, a practice normally adopted in the production of other distilled beverages such as whisky, cognac and rum, was first proposed by Novaes (1994) with the aim of obtaining a lighter distillate for later ageing. Recently, the bidistillation technique has been implemented to better standardise the profile of secondary organic compounds in sugar cane spirits produced in stills and to obtain a sensorially differentiated drink, with selectivity of the desired fractions, a reduction in volatile acidity, copper and aldehyde content (ROTA, 2008) and also to reduce or eliminate undesirable compounds such as methanol, furfural and ethyl carbamate (FORLIN, 2005).

This process consists of carrying out two successive distillations, which can be done in the same still or in different stills. Bi-distillation produces a more standardised spirit, with lower acidity and more pleasant sensory characteristics (ROTA, 2008). Currently, there are dozens of commercial brands of double-distilled spirits on the market, but in general, this practice is not yet adopted by small Brazilian producers. Sugarcane spirits are usually obtained in a single distillation, without any volatile fractions other than those contained in the head and tail being separated during the process (NOGUEIRA & VENTURINI-FILHO, 2005).

REDISTILLATION

In order to achieve similar goals, some producers have adopted the practice of re-distilling cachaças considered to be of poor quality. In this case, "ready-made" spirits are diluted to 25% alcohol and distilled again. In this second distillation, the head and tail fractions are separated and only the heart fraction is used. The brandy obtained in this way can, like double-distilled brandy, be bottled directly or go through an ageing process.

AGING

Ageing is undoubtedly an important stage in improving the quality attributes of cachaça, thus obtaining a product with quality compatible with the demands of the foreign market, as well as conquering the domestic market, since it is already evident that there is an increase in the consumption of artisanal cachaças with differentiated quality (FARIA, et al, 2003).

The natural ageing process of beverages consists of storing them in wooden barrels for a certain period of time, an action that produces results in the chemical composition, aroma, flavour and colour of the beverage and, therefore, its sensory quality. Numerous transformations take place during ageing, including reactions between secondary compounds from distillation, direct extraction of wood components, decomposition of some wood macromolecules (lignin, cellulose and hemicellulose) and their incorporation into the drink, as well as reactions of wood compounds with the original components of the distillate (PIGGOTT et al., 1989).

ADDED SUGAR

The addition of sugar (sucrose) to freshly distilled cachaça is quite common, especially among producers who obtain their drink by column distillation, and by bottlers who make their brandy from a mixture of other spirits usually bought from small producers (AQUINO, 2009).

The main objectives of this practice are: to soften the dry, burning flavour and characteristic aggressiveness of newly distilled products, to homogenise and standardise cachaça blends resulting from the mixing of spirits from different origins (a practice widely used by standardisers), and possibly to sensorially mask the high acidity of cachaças that have suffered bacterial contamination problems during the alcoholic fermentation of the mash, or errors during distillation (AQUINO, 2009).

Such deposits can form flakes in the cachaça, devaluing its commercial value.

LEGISLATION

According to Brazilian legislation (Normative Instruction No. 13 of 29 June 2005), sugar cane spirit is defined as "a drink with an alcoholic strength of 38 to 54% by volume, at twenty degrees Celsius, obtained from the simple alcoholic distillate of sugar cane or from the distillation of the fermented must of sugar cane juice, with the possibility of adding sugars of up to six grams per litre, expressed as sucrose".

Cachaça is the typical and exclusive name for sugarcane spirit produced in Brazil, with an alcoholic strength of 38 to 48% by volume at twenty degrees Celsius, obtained by distilling the fermented must of sugarcane juice with peculiar sensory characteristics, and may have added sugars of up to six grams per litre, expressed as sucrose. When sugar is added in excess of six grams and up to thirty grams per litre, the drink must bear the designation "sweetened" on the label (BRASIL, 2005).

Normative Instruction No. 13 determines the Chemical Composition and Quality Requirements for Cane Spirit and Cachaça. "The Coefficient of Congeners (components volatile "non-alcohol", or volatile substances "non-alcohol", or secondary components "non-alcohol", or volatile impurities "non-alcohol") is the sum of:

Volatile acidity (expressed as acetic acid, in mg/100 mL of anhydrous alcohol);

A Aldehydes (expressed as acetaldehyde, in mg/100 mL of anhydrous alcohol);

IS Total esters (expressed in ethyl acetate, in mg/100 mL of anhydrous alcohol);

Á Higher alcohols (expressed as the sum of n-propyl alcohol, isobutyl alcohol and isoamyl alcohols, in mg/100 mL of anhydrous alcohol);

F Furfural + hydroxymethylfurfural (expressed in mg/100 mL of anhydrous alcohol).

The Congeners Coefficient for Aguardente de cana and Cachaça may not be less than 200 mg per 100 mL and may not exceed 650 mg per 100 mL of anhydrous alcohol, and the following limits must be observed:

A Volatile acidity: maximum 150;

A Total aldehydes: maximum 30;

IS Total esters: maximum 200;

S Sum of isobutyl, isoamyl and n-propyl alcohols: maximum 360;

S Sum of furfural and hydroxymethylfurfural: maximum 5.

The same Normative Instruction sets limits for Inorganic Contaminants for both sugar cane spirit and Cachaça. These drinks must have:

C Copper in quantities not exceeding 5mg/L;

C Lead in quantities not exceeding 200μg/L **;

A Arsenic in quantities not exceeding 100μg/L **.

C Ethyl carbamate not exceeding 150μg/L *.

1 value to be required from June 2010 (BRASIL, 2005).

2 * values required as of June 2008 (BRASIL, 2005).

QUALITY CONTROL

Despite the tradition and economic importance of this drink, the country's brandy production chain is not technologically homogeneous, and there is a quest to develop technologies to improve and control the quality and standardisation of the drink. Improving the quality and standardisation of brandy and cachaça is essential if the drink is to meet international standards and be accepted by the foreign market, providing the conditions for opening and maintaining the market.

the export market. In addition, it would provide acceptance in the domestic market by the higher purchasing power classes, who demand a good quality drink (MIRANDA, 2007).

Faced with these new demands, producers are starting to worry about adding more value to the drink, both in terms of quality control and appearance, offering more sophisticated packaging, with the aim of being included in the country's best drinking spots alongside other well-established spirits (MIRANDA, 2007).

This makes it essential to control the quality of cachaças from both a sensory and physico-chemical point of view.

To this end, some analyses are essential to maintain and control the quality of the product, also ensuring that it remains within the parameters set by our legislation, such as: Copper, Higher Alcohols, Ethyl Acetate, Acetaldehyde, Methanol, Volatile Acidity, Relative Density and Real Alcohol Degree.

BIBLIOGRAPHICAL REFERENCES

AQUINO, F. W. B, **Dextran in sugars and cane spirits**. Thesis (PhD) - University of São Paulo, São Carlos, 2009.

BIZELLI, L. C.; RIBEIRO, C. A. F.; NOVAES, F. V. Double distillation of sugar cane spirit: total acidity and copper content. **Scientia Agrícola**, Piracicaba, v. 57, n. 4, p. 623627, 2000.

BRAZIL, Normative Instruction No. 13 of 30 June 2005. Approves the Technical Regulation for Setting the Standards of Identity and Quality for Aguardente de Cana and Cachaça. Ministry of Agriculture, Livestock and Supply. **Federal Official Gazette,** DF, 30 June 2005, Section 1.

CANTÃO, F. O. **"Physical-chemical analyses and evaluation of the presence of copper in sugar cane spirits using aluminosilicates"** - Master's thesis, UFLA, 2006.

FARIA, J. B; LOYOLA, E. ; LOPES, M.G. ; DUFOUR, J.P. . **Cachaça, Pisco and Tequila . In: LEA, A G H; PIGGOTT J R. (Org.). Fermented Beverage Production** 2nd. New York: Kluwer Academic / Plenium Publishers, 2003, v. , p. 335-363.

FARIA, J. B.; FERREIRA, V.; LOPEZ, R.; CACHO, J. **The sensory characteristic defect** of "cachaça" distilled in absence of copper. Revista Alimentos e Nutrição, Araraquara, v. 14, n. 1, p. 1-7, 2003

FARIA, J. B. **Determination of the compounds responsible for the sensory defect of cane spirits distilled**

in the absence of copper. 2000. 94 f. Thesis (Livre- Docência). Faculty of Pharmaceutical Sciences - UNESP - Araraquara- SP.

FARIA, J.B.; FRANCO, D.W.; CARDELLO, H.M.A.B.; BÔSCOLO M.; LIMA NETO, B.S. **Sensory evaluation of sugar cane brandy (Saccharum officinarum I.) during ageing in oak barrels (Quercus sp).** Ver. Brás. Anal. Alim., v.1, p.7 - 14, 1995.

FORLIN, F.J. **Maturation of sugar cane spirit composed with oak wood extract in polyethylene terephthalate (PET) packaging.** Thesis (doctorate) - State University of Campinas. Faculty of Food Engineering, 2005.

LIMA, U. A. **"Study of the main factors affecting the components of the non-alcohol coefficient of sugar cane spirits".** 1964. 141p. Thesis (full professor). Luiz de Queiroz College of Agriculture, Piracicaba: University of São Paulo.

LIMA, U.A. Aguardente. In: AQUARONE, E.; LIMA, U.A.; BORZANI, W. **Alimentos e bebidas produzidos por fermentação (Série biotecnologia)**. São Paulo: Edgard Blucher, 1983. v.5, cap.4, p. 79 - 102.

LIMA, U. A. **Manufacture in small distilleries**. Piracicaba: Fundação Estudos Agrários "Luiz de Queiroz", 1999. 187 p.

MIRANDA, M. B; MARTINS, N. G. S; BELLUCO, J. H; ALCARDE, A. R. **"Chemical quality of Brazilian cachaças and brandies"** Ciências Tecnologia de Alimentos, Campinas, Oct.-Dec. 2007

NÓBREGA, I.C.C. - **"Analysis of the volatile compounds of sugar cane spirit by dynamic headspace concentration and gas chromatography-mass spectrometry"** Ciências e Tecnologia de alimentos, Campinas, 23(2), p.210 - 216, May-August 2003.

NOGUEIRA, A.M.P.; FILHO, W.G.V. - **"AGUARDENTE DE CANA"** - UNESP Campus De Botucatu Faculdade De Ciências Agronômicas, 2005.

NOVAES, F.V. **"Production and quality of sugar cane brandy"**. Piracicaba: ESALQ, 1995. 27p. Handout.

NOVAES, F. V. **Basic notions on distillation theory.** Piracicaba: ESALQ/Depto de Ciência e Tecnologia Agroindustrial, 1994. 22p

PIGGOTT, J. R. Ç.; SHARP, R. Ç.; DUNCAN, R. E. B. **The Science and Technology of Whiskies**. New York: Longman Scientific & Technical, 1989. 410 p.

ROTA, M. B. **Effect of the bidistillation process on the sensory quality of cachaça.** 2008. 80p. Dissertation (Master's in Food Science) - Universidade Estadual Paulista "Júlio de Mesquita Filho". Faculty of Pharmaceutical Sciences, Department of Food and Nutrition, Araraquara Campus. 2008.

YOKAYA, F. **"Fabricação da aguardente de cana"** (Industrial fermentations series, n°2). Campinas, SP: Fundação Tropical de Pesquisa e Tecnologia "André Tosello", 1995. 93 p.

CHAPTER 2

Comparative study (sensory and physico-chemical) between the traditional process and redistillation to obtain cachaça

Study (sensory and physical-chemical) comparing the traditional process and redistillation to obtain "cachaça"

SUMMARY:

Based on the positive effect that the bi-distillation process has on reducing sensory defects and standardising cachaça and other distilled beverages that normally use this process to obtain them, the aim of this study was to see if the re-distillation of traditionally obtained cachaça samples could improve the sensory quality of this drink.

To this end, two samples of cachaça were obtained from the same batch of sugar cane spirit: 1 - the so-called traditional sample (obtained by diluting the alcohol content of the batch mentioned above) and; 2 - the so-called redistilled sample (obtained by redistilling and diluting the same batch). The results obtained in the acceptance tests, purchasing attitude and physico-chemical determinations reveal the positive role of redistillation in the sensory quality of cachaça and point to the reduction in volatile acidity as at least co-responsible for the effect observed.

Keywords: Cachaça, Redistillation, Sensory Analysis, Volatile Acidity.

INTRODUCTION

Sugarcane brandy is the second most consumed alcoholic drink in Brazil, with production estimated at around 1.2 billion litres/year. When the distilled spirits market is considered exclusively, it ranks first in the country (87% of the market). In 2009, 10.83 million litres were exported, generating revenues of US$15.58 million. Among the main destination markets are Germany, the United States, Portugal and France (ALVES, 2009).

However, the volume of cachaça exports still represents less than 1% of the total produced (ALVES, 2009).

There are currently more than 40,000 producers (4,000 brands), with micro-enterprises accounting for 99% of all producers. Their agricultural activities include the production of corn, beans, coffee and milk, among others, and the production of cachaça. The cachaça sector is responsible for generating more than 600,000 direct and indirect jobs (ALVES, 2009).

Despite the tradition and economic importance of this drink, the country's brandy production chain is not technologically homogeneous, and there is a quest to develop technologies to improve and control the quality and standardisation of the drink. Improving the quality and standardisation of brandy and cachaça is essential if the drink is to meet international standards and be accepted by the foreign market, providing the conditions for opening up and maintaining the export market. In addition, it could also provide acceptance in the domestic market by the more affluent classes, who demand good quality drinks (MIRANDA, 2007).

In order to correct sensory defects in traditional cachaça samples, some producers have adopted the practice of redistilling ready-made cachaças (38-48% alcohol) after dilution to around 30% alcohol or less, with the aim of obtaining a product with more neutral (smooth) characteristics, similar to those obtained using the bidistillation process. In this second distillation, the initial (head) and final (tail) fractions are separated, resulting in a distillate similar to that obtained by the double distillation process (FRANCO, 2009).

Sensory analysis is a tool that can be used to measure, analyse and interpret the impact that the different characteristics of foods, drinks and materials have on the human sense organs, and thus determine how products are perceived. It is an important technique for evaluating the sensory quality of various products, as there is still no instrument capable of replacing the human senses in characterising and evaluating these products when they are consumed (JELLINEK, 1985 cited by MAÇATELLI, 2007).

Sensory tests, initially simpler, have evolved and become increasingly complex. At the same time, more advanced statistical programmes have been incorporated into sensory analysis, which has made it possible for sensory analysis to become a powerful tool to be used in the quality control of food, beverages, perfumes and other products (MAÇATELLI, 2007).

Based on the above, the aim of this study was to evaluate the effect of redistillation on the sensory quality of cachaça, comparing traditional and redistilled samples.

MATERIAL AND METHODS

MATERIAL

From a batch of traditional brandy with low sensory acceptance and an alcohol content of 47 per cent, the traditional sample was initially obtained by dilution to 40°GL, as is common practice with commercial samples.

The same batch as above, after dilution to 25% alcohol, was distilled again and the heart fraction, after correction to 40°GL, gave rise to the redistilled sample.

The brandy samples thus obtained were then subjected to acceptance tests, purchasing attitudes and physical-chemical and compositional determinations, with a view to verifying the effects of the redistillation process.

METHODS

SENSORY ANALYSIS

ACCEPTANCE TESTS

In order to make an initial comparison, the two samples of cachaça, traditional and redistilled, were then subjected to acceptance tests with regard to aroma, flavour, sweetness and overall impression, and to determine purchasing attitude.

To this end, 56 judges were selected through questionnaires (FIGURES 1 and 2), including students, teachers and staff from FCF-UNESP and UNIARA, over 21 years old and consumers of distilled beverages, who were interested in taking part in the tests. These judges were then given the cachaça samples for acceptance tests at individual booths in the Sensory Analysis Laboratory of the Food and Nutrition Department at FCF-UNESP and in the biology laboratory at Uniara. The samples were served randomly, monadically in transparent crystal glasses, coded with a

3-digit number and covered with watch glasses, which were removed at the time of the test. The scores were recorded on cards (FIGURE 3), with a structured 9-point hedonic scale, which also recorded what each judge liked best, least and their purchasing attitude towards the product evaluated.

Please complete the questionnaire with all the information requested. Name			
Gender: ( ) male	( ) female		
Age: years			
Teacher ( )	Employee ( )	Postgraduate student ( )	Undergraduate student ( )
Schooling: ( ) primary school		( ) high school	( ) incomplete higher education
	( ) university degree	( ) postgraduate (complete or incomplete)	
Telephone	Email		
Using the scale below, indicate how much you like alcoholic drinks:			
( ) I like it a lot ( ) I like it a lot ( ) I like it moderately ( ) I like it slightly ( ) Neither like nor dislike ( ) Slightly disgusted ( ) Moderately disgusted ( ) I really dislike it ( ) I dislike it a lot			
How often do you buy/consume alcoholic drinks?			
( ) 3 times/week or more ( ) 1 time/week ( ) 1 time/fortnight ( ) 1 time/month ( ) Does not consume			

FIGURE 1: Personal data questionnaire on alcohol consumption and taste

Please complete the questionnaire with all the information requested. Name	
Using the scale below, indicate how often you consume the different alcoholic drinks.	
	( ) Unaged cachaça
(4)3 times/week or more	( ) Aged cachaça
(3)1 time/week	( ) Unaged spirits (Vodka, Rum, Gin)
(2)1 time/fortnight	( ) Aged spirits (whisky, rum)
(1)1 time/month	( ) Caipirinha (Cachaça with lemon or any other fruit)
(0) Does not consume	( ) Caipirosca (Vodka with lemon or any other fruit)
	( ) Caipiríssima (Rum with lemon or any other fruit)
	( ) Fermented (Beer, Wine)
	( ) Drinks (Alexander, Silk Stocking, Blue Lagoon, etc)
	( ) Aperitifs (Martini, San Remy)
	( ) Liqueurs (Amarula, Frangelico, Amareto, etc.)
Using the scale below, indicate how much you like the alcoholic drinks you consume, according to **the previous question.**	
	( ) Unaged cachaça
(9) I like it a lot	( ) Aged cachaça
(8) I like it a lot	( ) Unaged spirits (Vodka, Rum, Gin)
(7) I like it moderately	( ) Aged spirits (whisky, rum)
(6) I like it slightly	( ) Caipirinha (Cachaça with lemon or any other fruit)
(5) Neither like nor dislike	( ) Caipirosca (Vodka with lemon or any other fruit)
(4) Slightly disgusted	( ) Caipiríssima (Rum with lemon or any other fruit)
(3) Moderate dislike	( ) Fermented (Beer, Wine)
(2) I really dislike it	( ) Drinks (Alexander, Silk Stocking, Blue Lagoon, etc)
(1) I really dislike it	( ) Aperitifs (Martini, San Remy)
	( ) Liqueurs (Amarula, Frangelico, Amareto, etc.)

FIGURE 2: Questionnaire on the frequency and taste of various alcoholic drinks

Name:	Date: / /
You are receiving a sample of CACHAÇA. Please rate the sample according to the attributes, using the scale below. Sample no:	

AROMA: Desgostei muitíssimo / Desgostei muito / Desgostei moderadamente / Desgostei ligeiramente / Nem gostei nem desgostei / Gostei ligeiramente / Gostei moderadamente / Gostei muito / Gostei muitíssimo

1——.——.——.——5——.——.——.——9

FLAVOUR: Desgostei muitíssimo / Desgostei muito / Desgostei moderadamente / Desgostei ligeiramente / Nem gostei nem desgostei / Gostei ligeiramente / Gostei moderadamente / Gostei muito / Gostei muitíssimo

1——.——.——.——5——.——.——.——9

SWEETNESS: Desgostei muitíssimo / Desgostei muito / Desgostei moderadamente / Desgostei ligeiramente / Nem gostei nem desgostei / Gostei ligeiramente / Gostei moderadamente / Gostei muito / Gostei muitíssimo

1——.——.——.——5——.——.——.——9

1 ---------------------------------5----------------------------------- 9

OVERALL IMPRESSION: Desgostei muitíssimo / Desgostei muito / Desgostei moderadamente / Desgostei ligeiramente / Nem gostei nem desgostei / Gostei ligeiramente / Gostei moderadamente / Gostei muito / Gostei muitíssimo

1——.——.——.——5——.——.——.——9

Indicate below what you liked best and what you liked least about this aged brandy:

+ I liked it: ____________________

- I liked it: ____________________

Tick below the degree of certainty with which you would or would not buy this aged brandy:

( ) Would certainly buy
( ) Probably would
( ) I'm not sure I'd buy it
( ) I probably **wouldn't** buy it
( ) I certainly **wouldn't** buy it

Comments: ____________________

FIGURE 3: Sheets, with a 9-point structured hybrid scale, which also recorded what each judge liked best, least and what their purchasing attitude was towards the product evaluated.

PHYSICO-CHEMICAL DETERMINATIONS

Samples number 1 (cachaça obtained traditionally) and number 2 (cachaça obtained by the redistillation process) were analysed:

THE COPPER CONTENT

A colourimetric method was used (BRASIL, 1986)

THE PROFILE OF VOLATILE COMPOUNDS

The concentrations of volatile compounds were determined using gas chromatography (EC, 2000).

VOLATILE ACIDITY

Using a titrimetric method (BRASIL, 1986).

THE ACTUAL ALCOHOLIC STRENGTH

Using a densimetric method (EC, 2000).

RESULTS AND DISCUSSION

After selecting the judges based on the questionnaire data, and selecting those who ticked between "I like alcoholic beverages a little" and "I like them a lot" (Figure 4) or who consumed alcoholic beverages at least once a month (Figure 5), a panel of 56 judges was established, 57 per cent of whom were male and 43 per cent female (Figure 6), the vast majority of whom were undergraduate students (67 per cent) and postgraduate students (21 per cent) (Figure 7).

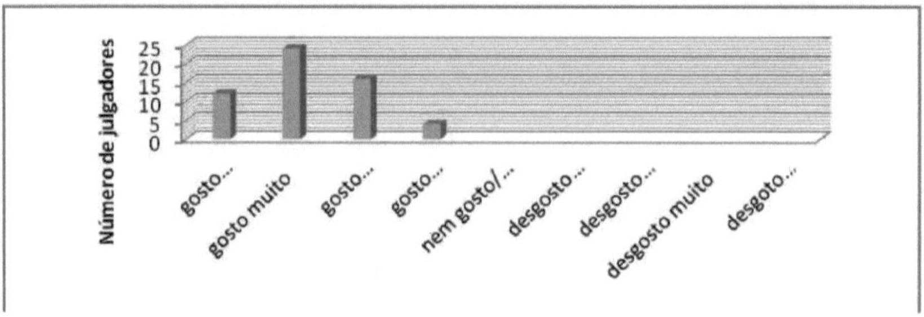

Figure 4: Graphical representation of the results of the selected judges in relation to how much they like or dislike alcoholic drinks.

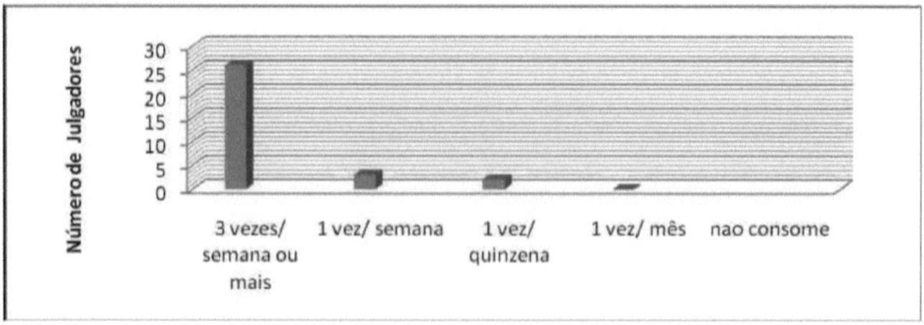

Figure 5: Graphical representation of the results of the selected judges in relation to the frequency of alcohol consumption.

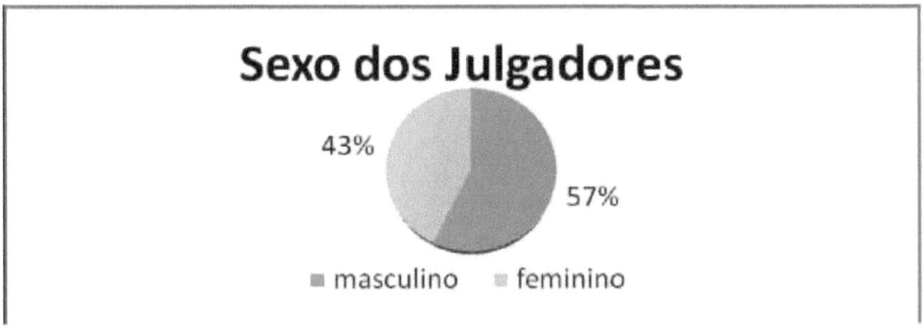

Figure 6: Characterisation of the judges in relation to gender.

Figure 7: Characterisation of the judges in relation to their profession.

With regard to education (Figure 8), it can be seen that the majority of judges have incomplete higher education (62%), followed by graduates (27%).

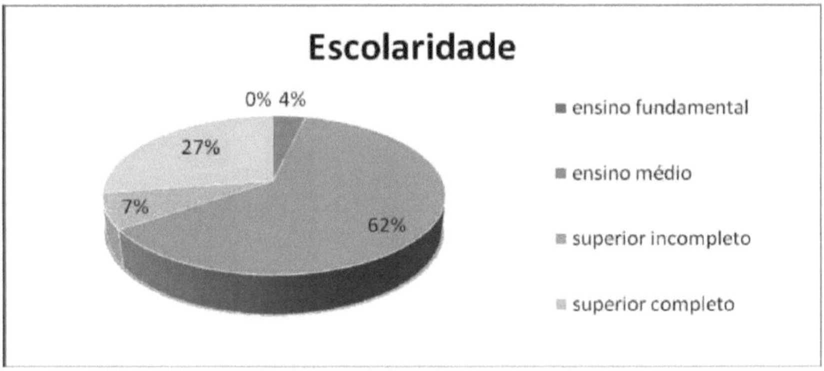

Figure 8: Characterisation of the judges in terms of education level.

In terms of age (Figure 9), there was a greater participation of young people aged between 18 and 22 (45 per cent), followed by a group aged between 23 and 26 (31 per cent).

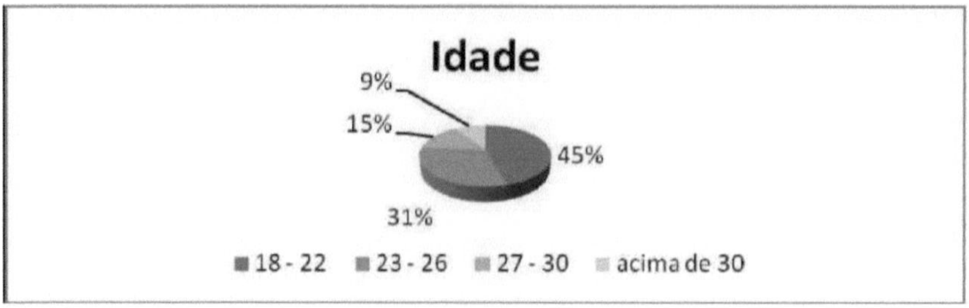

Figure 9: Characterisation of the judges in terms of age group.

ACCEPTANCE TESTS

The results of the acceptance tests in relation to aroma, flavour, sweetness and overall impression were subjected to analysis of variance (ANOVA) and Tukey's test and the results are shown in Table 1.

TABLE [1]: Averages 1 obtained in the acceptance test for traditional (1) and redistilled (2) cachaça samples.

SAMPLE	AROMA	FLAVOUR	SWEETNESS	IMP. GLOBAL
1	5,8071 ^{A}	4,9304 ^{A}	4,6911 ^{A}	4,9089 ^{A}
2	6,3214 ^{B}	5.5232^{A}	5.3446^{A}	5.6429 ^{B}
DMS	0.4913	0.6412	0.6592	0.6612

[1] Averages with the same letter in the same column did not differ (p<0, 05), n=56 tasters.

The results showed a significant difference ($p<0.05$) between the flavour and overall impression attributes.

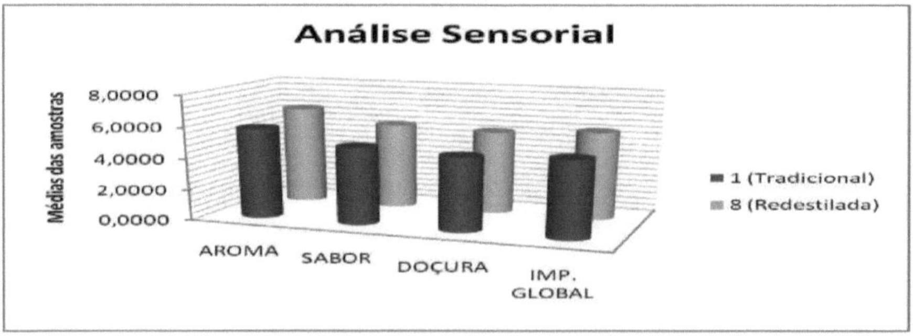

Figure 10: Graphical representation of the sensory analysis

The sensory analysis showed a significant difference for the aroma and overall impression parameters, proving the influence of redistillation on the sensory quality of cachaça.

Although the samples did not differ significantly in terms of flavour and sweetness, the averages

(Figure 10) also show a trend towards greater acceptance when comparing the traditional sample with the redistilled one. Similarly, the results obtained in relation to purchasing attitude (Figure 11) show that the redistilled sample not only obtained greater acceptance, but also a higher level of purchase compared to the traditional sample, thus confirming the positive effect of the redistillation process on the sensory quality of cachaça.

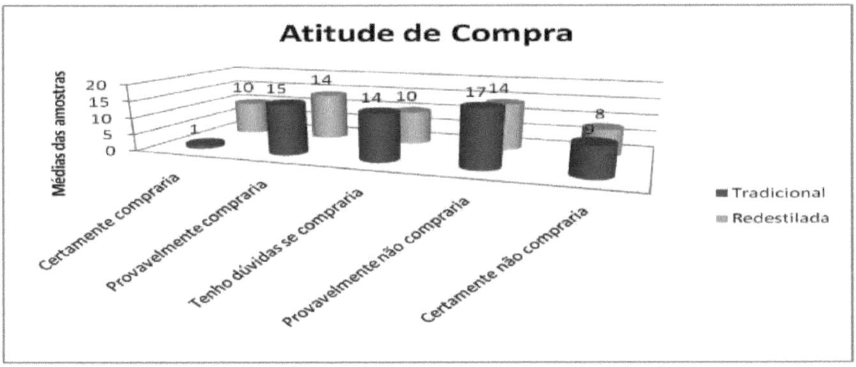

Figure 11: Graphical representation of the purchasing attitude of the samples analysed.

RESULTS OF PHYSICO-CHEMICAL DETERMINATIONS

Table 2 shows the results of the levels of copper, higher alcohols, ethyl acetate, acetaldehyde, methanol, volatile acidity, density and real alcoholic strength present in the traditional and redistilled samples.

Table 2 - Content of some compounds in the samples

Parameters	Sample 1 (traditional)	Sample 2 (redistilled)	Unit
Copper	0,50	0,60	mg/L
Higher alcohols	254,0	266,1	mg/100 mL p.a.
Ethyl Acetate	5,1	8,8	mg/100 mL p.a.
Acetaldehyde	3,4	5,7	mg/100 mL p.a.
Methanol	4,1	4,6	mg/100 mL p.a.
Volatile acidity	28,7	4,4	mg acetic acid/100 mL a.a.
Real alcoholic strength	36,55	38,01	°GL

The volatile acidity parameter (Figure 12) was the one that showed significant differences between the samples, with sample 1 (traditional) having the highest value (28.7 mg acetic acid/100 mL a.a.), much higher than sample 2 (redistilled) (4.4 mg acetic acid/100 mL a.a.).

In Figure 12, the aforementioned reduction can be better observed and is certainly directly related to the better acceptance of the redistilled cachaça samples, given the recognised negative effect of this parameter on the sensory quality of cachaça.

The fact that no other significant differences were observed between the other parameters evaluated does not rule out the possibility that other changes due to the effect of redistillation may also have influenced the sensory results obtained.

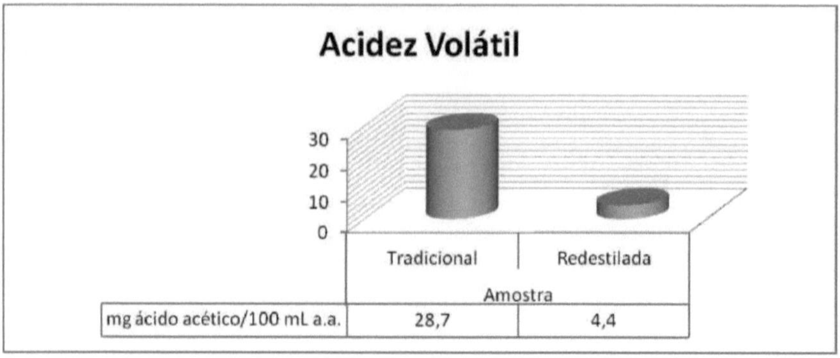

Figure 12: Graphical representation of the values obtained for the volatile acidity parameter.

CONCLUSIONS

The results obtained in this study show that the process of redistilling cachaça had a positive effect on its acceptance, since the redistilled samples always showed higher averages than the corresponding samples obtained traditionally, indicating that this is a valid option for correcting any defects in cachaça obtained under unsuitable conditions.

The results of the physico-chemical analyses reinforce the results obtained in the sensory analysis, relating the better acceptance scores of the redistilled samples to the reduction in volatile acidity, a factor that is known to be important for the sensory quality of cachaça.

ACKNOWLEDGEMENTS

To Capes for the financial support.

MAPA (Ministry of Agriculture, Livestock and Supply) for the physical and chemical analyses.

REFERENCES

BRAZIL. Ministry of Agriculture. Ordinance no. 76 of 26 November 1986. Provides for the analytical methods of beverages and vinegar. **Official Gazette of the Federative Republic of Brazil**, Brasília, 28 Nov. 1986. Section 1, pt. 2.

BRAZIL, Normative Instruction No. 13 of 30 June 2005. Approves the Technical Regulation for Setting the Standards of Identity and Quality for Aguardente de Cana and Cachaça. Ministry of Agriculture, Livestock and Supply. **Federal Official Gazette,** DF, 30 June 2005, Section 1.

CARDELLO, H.M.A.B.; FARIA J.B.-"Quantitative descriptive analysis of sugar cane spirit during ageing in oak barrels *(Quercus alba* L.)" - **Ciência e Tecnologia de Alimentos** - v. 18 n. 2 Campinas May/Jul. 2000.

FRANCO, A. C.; ROTA, M. B.; FARIA, J. B. Redistillation of Cachaça - **Alimentos e Nutrição** - v. 20, n.2, p. 331-334, 2009.

MAÇATELLI, M. **"Determining the Sensory Profile of Commercial Brands of Cachaça"** 2007. 126p. Dissertation (Master's Degree in Food Science) - Universidade Estadual Paulista "Júlio de Mesquita Filho". Faculty of Pharmaceutical Sciences, Department of Food and Nutrition, Araraquara Campus. 2007.

MIRANDA, M. B; MARTINS, N. G. S; BELLUCO, J. H; ALCARDE, A. R. **"Chemical quality of Brazilian cachaças and brandies"** Ciências Tecnologia de Alimentos, Campinas, Oct.-Dec. 2007

COMMISSION REGULATION (EC) No 2870/2000 of 19 December 2000 laying down Community reference methods for the analysis of spirit drinks THE COMMISSION OF THE EUROPEAN COMMUNITIES, **Official Journal of the European Union.**

ELECTRONIC REFERENCE

ALVES, M. | **Mercado de Cachaça** - Envasador Magazine - Issue 40 - Supplier Guide Mon, 02 March 2009 00:00 - Available at: http://www.ibrac.net. Accessed on: 15/09/2010.

CHAPTER 3

Effect of added sugar on the sensory quality of traditionally produced and redistilled cachaça*

Vitor Rocha dos Santos†

João Bosco FARIA**

Paper published in the Food and Nutrition Journal of the Faculty of Pharmaceutical Sciences, UNESP - Araraquara, ISSN: 0103-4235

Alim. Nutr., Araraquara v. 22, n. 3, p. 489-497, jul./set. 2011

Effect of sugar in the sensory quality of samples of "cachaça" traditionally distilled and redistilled

SUMMARY:

Sugarcane spirit is the second most consumed alcoholic drink in Brazil, with production estimated at around 1.2 billion litres per year. In 2009, 10.83 million litres were exported, generating revenues of US$15.58 million. Among the main destination markets are Germany, the United States, Portugal and France (ALVES, 2009).

The addition of sugar has been used as a way of standardising the taste of some distilled beverages in order to reduce the taste of ethanol and any sensory defects (CARDELLO and FARIA, 2000).

The aim of this study was to assess the effect of adding different concentrations of sugar on the sensory quality of cachaça, comparing traditionally produced and redistilled samples.

From the same batch of traditionally produced brandy, which was considered to be poorly accepted, seven samples were initially obtained, containing increasing concentrations of sugar, which were then diluted to 40°GL. Subsequently, a volume of the same batch was redistilled to obtain seven more samples, also containing increasing concentrations of sugar and also diluted to 40° GL.

The brandy samples thus obtained were then subjected to acceptance tests in relation to the attributes aroma, flavour, sweetness, overall impression and purchasing attitude.

The results showed that the practice of adding sugar to cachaça undoubtedly has an influence on consumer acceptance of the final product, minimising sensory defects related to inadequate production techniques, as well as that the redistillation process can improve the acceptance of spirits with low acceptance.

Keywords: Cachaça, redistillation, sensory analysis.

INTRODUCTION

Sugarcane spirit is the second most consumed alcoholic drink in Brazil, with production estimated at around 1.2 billion litres/year . When the distilled spirits market is considered

---

* Part of the first author's Master's Dissertation, in 2010, in the Food Science Course, Department of Food and Nutrition - Faculty of Pharmaceutical Sciences - UNESP - 14801-902 - Araraquara - SP - Brazil. Work carried out with financial support from CAPES, in the form of a Master's scholarship.

† Department of Food and Nutrition - Faculty of Pharmaceutical Sciences - UNESP - 14801-902 - Araraquara - SP - Brazil

exclusively, it ranks first in the country (87% of the market). In 2009, 10.83 million litres were exported, generating revenues of US$15.58 million. Among the main destination markets are Germany, the United States, Portugal and France (ALVES, 2009).

However, the volume of cachaça exports still represents less than 1% of the total produced (ALVES, 2009).

Exports are low mainly due to the fact that Brazilian brandy doesn't have a high quality standard, mainly due to the technical conditions of this agro-industry, which, with a few exceptions, leave a lot to be desired, mainly due to the lack of knowledge of the process and effective control of its quality, which can guarantee a higher quality product, as well as better industrial yields.

Faced with new market demands, producers are starting to worry about adding value to the drink, investing in new technologies, improved quality control and more sophisticated packaging, with the aim of placing cachaça in better consumption points, alongside other distilled drinks that are already well-established in the country and around the world.

The addition of sugar has been used as a way of standardising the taste of distilled beverages, attenuating (reducing) the taste of ethanol and any sensory defects (CARDELLO and FARIA, 2000).

In the case of cachaça, Brazilian legislation allows the addition of up to 6g of sugar per litre, as a normal procedure for standardising this drink. However, if the term "sweetened" is stated on the label, the addition of sugar can reach up to 30g/L (BRASIL, 2005). In this case, it is worth questioning whether this amount is not excessively high and therefore represents a way of masking the flavour of poorly processed cachaças.

With a view to achieving similar goals in terms of the final quality of the product, some producers have also adopted the practice of re-distilling cachaças considered to be of poor quality. In this case, "ready-made" spirits, after dilution to around 25 per cent alcohol, are distilled again, the head and tail portions separated and only the heart portion used, which can then be bottled directly or go through the ageing process.

Sensory analysis is a tool that can be used to measure, analyse and interpret the impact that the different characteristics of foods, drinks and materials have on the human sense organs, and thus determine how products are perceived. It is an important technique for evaluating the sensory quality of various products, as there is still no instrument capable of replacing the human senses in characterising and evaluating these products when they are consumed (JELLINEK, 1985 cited by MAÇATELLI, 2007).

Sensory tests, initially simpler, have evolved and become increasingly complex. At the same time, more advanced statistical programmes have been incorporated into sensory analysis, which has made it possible for sensory analysis to become a powerful tool to be used in the quality control of food, beverages, perfumes and other products (MAÇATELLI, 2007).

Based on the above, the aim of this study was to assess the effect of adding sugar and redistillation on the sensory quality of cachaça, comparing traditional samples and those redistilled under similar conditions, but containing different concentrations of sugar.

MATERIAL AND METHODS

MATERIAL

From a batch of traditional brandy with low sensory acceptance and an alcohol content of 47 per cent, seven samples were initially obtained, containing increasing concentrations of sugar, and diluted to 40°GL. According to the most common practice observed in commercial samples

Sample

1 - Traditional (Control)
2 - Traditional with 0.6% added sugar
3 - Traditional with 1.0% added sugar
4 - Traditional with 1.5% added sugar
5 - Traditional with 2.0% added sugar

6 - Traditional with 2.5% added sugar
7 - Traditional with 3.0% added sugar

After dilution to 25% alcohol, the same batch was distilled again and the heart fraction obtained in this process was used to obtain the seven other samples listed below, also corrected to 40°GL.

8 - Redistilled (Control)

9 - Redistilled with 0.6 % sugar
10 - Redistilled with 1.0% added sugar
11 - Redistilled with 1.5% added sugar
12 - Redistilled with 2.0% added sugar
13 - Redistilled with 2.5% added sugar
14 - Redistilled with 3.0% added sugar

To obtain the sugar concentrations present in each sample, dilution processes were carried out, always starting from a mother solution with the highest sugar concentration (3%).

The 14 brandy samples thus obtained were then subjected to acceptance tests, purchasing attitudes and physico-chemical and compositional determinations in order to compare the effects of the redistillation process and the addition of sugar.

METHODS

SENSORY ANALYSIS

ACCEPTANCE TESTS

In order to make an initial comparison, the 14 cachaça samples were then subjected to acceptance tests with regard to aroma, flavour, sweetness and overall impression, and to determine purchasing attitude.

To this end, 56 judges were selected through questionnaires (FIGURES 1 and 2), including students, teachers and staff from FCF-UNESP and UNIARA, over 21 years old and consumers of distilled beverages, who showed interest in taking part in the tests. These judges were then given the cachaça samples for acceptance tests in individual booths in the Sensory Analysis

Laboratory of the Food and Nutrition Department at FCF-UNESP and in the biology laboratory at Uniara. The samples were served randomly, monadically in transparent crystal glasses, coded with a 3-digit number and covered with watch glasses, which were removed at the time of the test. The scores were recorded on cards (FIGURE 3), with a structured 9-point hedonic scale, which also recorded what each judge liked best, least and their purchasing attitude towards the product evaluated.

Please complete the questionnaire with all the information requested. Name			
Gender:	( ) male	( ) female	
Age:	years		
Teacher ( )	Employee ( )	Postgraduate student ( )	Undergraduate student ( )
Schooling: ( ) primary school		( ) high school	( ) incomplete higher education
	( ) university degree	( ) postgraduate (complete or incomplete)	
Telephone	Email		

Using the scale below, indicate how much you like alcoholic drinks:
( ) I like it a lot
( ) I like it a lot
( ) I like it moderately
( ) I like it slightly
( ) Neither like nor dislike
( ) Slightly disgusted
( ) Moderately disgusted
( ) I really dislike it
( ) I dislike it a lot

How often do you buy/consume alcohol?
( ) 3 times/week or more
( ) 1 time/week
( ) 1 time/fortnight
( ) 1 time/month
( ) Does not consume

FIGURA 1: Questionnaire on personal data, taste and alcohol consumption

Please complete the questionnaire with all the information requested. Name

Using the scale below, indicate how often you consume the different alcoholic drinks.

	( ) Unaged cachaça
(4)3 times/week or more	( ) Aged cachaça
(3)1 time/week	( ) Unaged spirits (Vodka, Rum, Gin)
(2)1 time/fortnight	( ) Aged spirits (whisky, rum)
(1)1 time/month	( ) Caipirinha (Cachaça with lemon or any other fruit)
(0) Does not consume	( ) Caipirosca (Vodka with lemon or any other fruit)
	( ) Caipiríssima (Rum with lemon or any other fruit)
	( ) Fermented (Beer, Wine)
	( ) Drinks (Alexander, Silk Stocking, Blue Lagoon, etc)
	( ) Aperitifs (Martini, San Remy)
	( ) Liqueurs (Amarula, Frangelico, Amareto, etc.)

Using the scale below, indicate how much you like the alcoholic drinks you consume, according to **the previous question.**

	( ) Unaged cachaça
(9) I like it a lot	( ) Aged cachaça
(8) I like it a lot	( ) Unaged spirits (Vodka, Rum, Gin)
(7) I like it moderately	( ) Aged spirits (whisky, rum)
(6) I like it slightly	( ) Caipirinha (Cachaça with lemon or any other fruit)
(5) Neither like nor dislike	( ) Caipirosca (Vodka with lemon or any other fruit)
(4) Slightly disgusted	( ) Caipiríssima (Rum with lemon or any other fruit)
(3) Moderate dislike	( ) Fermented (Beer, Wine)
(2) I really dislike it	( ) Drinks (Alexander, Silk Stocking, Blue Lagoon, etc)
(1) I really dislike it	( ) Aperitifs (Martini, San Remy)
	( ) Liqueurs (Amarula, Frangelico, Amareto, etc.)

FIGURA 2: Questionnaire on the frequency and taste of various alcoholic drinks

Name: Date: / /

You are receiving a sample of CACHAÇA. Please rate the sample according to the attributes, using the scale below.

Sample no: ______________

Desgostei muitíssimo / Desgostei muito / Desgostei moderadamente / Desgostei ligeiramente / Nem gostei nem desgostei / Gostei ligeiramente / Gostei moderadamente / Gostei muito / Gostei muitíssimo

AROMA: 1——.——.——.——5——.——.——.——9

Desgostei muitíssimo / Desgostei muito / Desgostei moderadamente / Desgostei ligeiramente / Nem gostei nem desgostei / Gostei ligeiramente / Gostei moderadamente / Gostei muito / Gostei muitíssimo

FLAVOUR: 1——.——.——.——5——.——.——.——9

Desgostei muitíssimo / Desgostei muito / Desgostei moderadamente / Desgostei ligeiramente / Nem gostei nem desgostei / Gostei ligeiramente / Gostei moderadamente / Gostei muito / Gostei muitíssimo

SWEETNESS: 1——.——.——.——5——.——.——.——9

Desgostei muitíssimo / Desgostei muito / Desgostei moderadamente / Desgostei ligeiramente / Nem gostei nem desgostei / Gostei ligeiramente / Gostei moderadamente / Gostei muito / Gostei muitíssimo

OVERALL IMPRESSION: 1——.——.——.——5——.——.——.——9

Indicate below what you liked best and what you liked least about this aged brandy:

+ I liked it: ______________________________

- I liked it: ______________________________

Tick below the degree of certainty with which you would or would not buy this aged brandy:

( ) I would certainly buy
( ) I would probably buy
( ) I'm not sure I'd buy
( ) I probably **wouldn't** buy
( ) I certainly **wouldn't** buy

Comments: ______________________________

FIGURA 3: Sheets, with a structured 9-point hybrid scale, where each judge also recorded what they liked best, what they liked least and what their purchasing attitude was towards the product being evaluated.

FIGURA 4:

PHYSICO-CHEMICAL DETERMINATIONS

Samples number 1 (cachaça obtained traditionally, without added sugar) and number 8 (cachaça obtained by the redistillation process, without added sugar) were analysed to compare the results in terms of the influence of the processes.

THE COPPER CONTENT

A colourimetric method was used (BRASIL, 1986)

THE PROFILE OF VOLATILE COMPOUNDS

The concentrations of volatile compounds were determined using gas chromatography (EC, 2000).

VOLATILE ACIDITY

Using a titrimetric method (BRASIL, 1986).

THE ACTUAL ALCOHOLIC STRENGTH

Using a densimetric method (EC, 2000).

RESULTS AND DISCUSSION

After selecting the judges based on the questionnaire data, and selecting those who ticked between "I like alcoholic beverages a little" and "I like them a lot" (Figure 4) or who consumed alcoholic beverages at least once a month (Figure 5), a pool of 56 judges was established, 57 per cent of whom were male and 43 per cent female (Figure 6), the vast majority of whom were undergraduate students (67 per cent) and postgraduate students (21 per cent) (Figure 7).

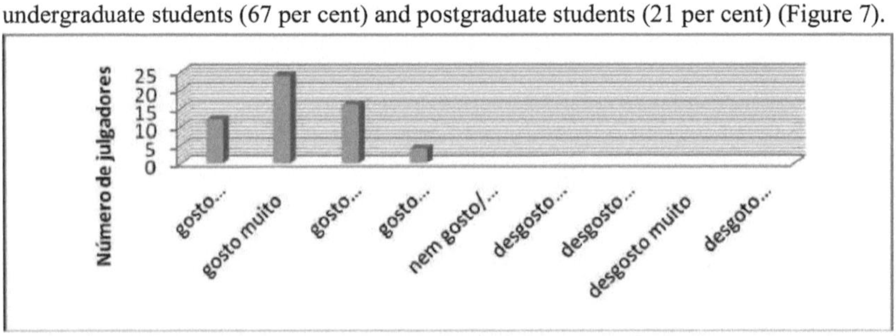

Figure 4: Graphical representation of the results of the selected tasters in relation to how much they like or dislike alcoholic drinks.

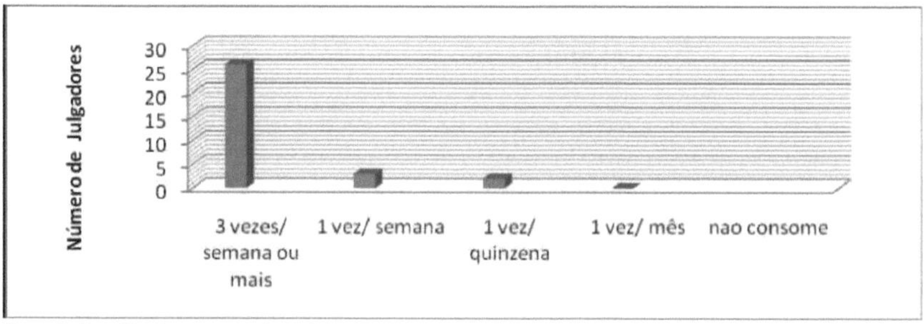

Figure 5: Graphical representation of the results of the selected judges in relation to the frequency of alcohol consumption.

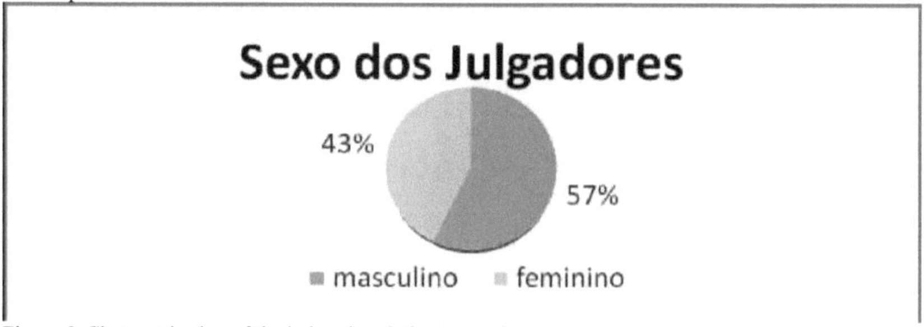

Figure 6: Characterisation of the judges in relation to gender.

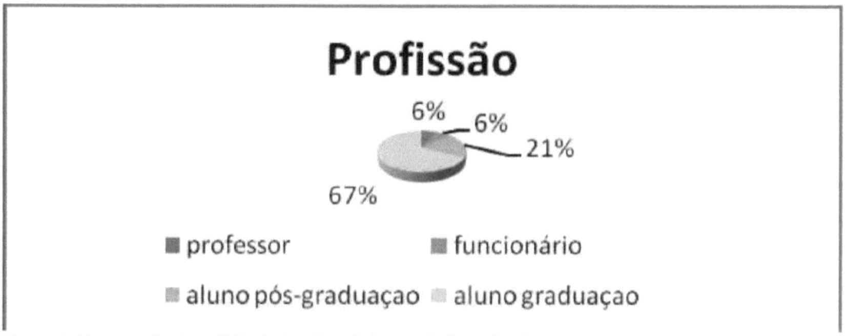

Figure 7: Characterisation of the judges in relation to their profession.

In terms of education (Figure 8), it can be seen that the majority of judges have incomplete higher education (62%), followed by graduates (27%).

Figure 8: Characterisation of the judges in terms of education level.

In terms of age (Figure 9), there was a greater participation of young people aged between 18 and 22 (45 per cent), followed by a group aged between 23 and 26 (31 per cent).

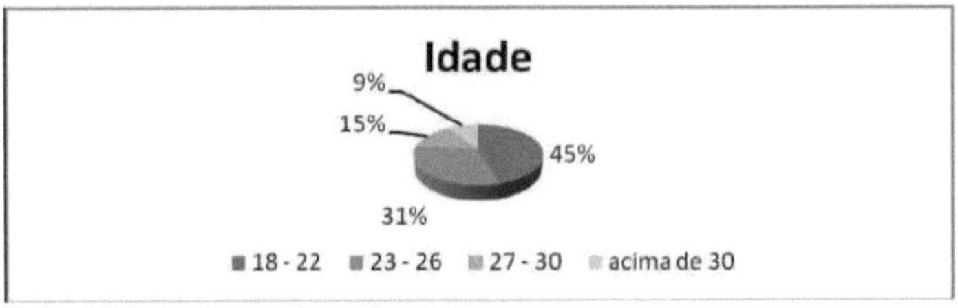

Figure 9: Characterisation of the judges in terms of age group.

ACCEPTANCE TEST

The results of the acceptance tests in relation to aroma, flavour, sweetness and overall impression were then subjected to analysis of variance (ANOVA) and Tukey's test, as shown in Table 1.

TABLE [1]: Averages 1 obtained in the acceptance test of traditional and redistilled cachaça samples, varying the sugar concentrations.

SAMPLE	AROMA	FLAVOUR	SWEETNESS	IMP. GLOBAL
1	5,8071 ^{B}	4,9304 ^{E}	4,6911 ^{C}	4,9089 ^{C}
2	6.3214 ^{BA}	5.6732 ^{BDEC}	5.5857 ^{BAC}	5.8000 ^{BAC}
3	5.9786 ^{BA}	5.4679 ^{DE}	5.3589 ^{BC}	5.4821 ^{BC}
4	6.6018 ^{BA}	6.1714 ^{BDAC}	5.8893 ^{BA}	6.2304 ^{BA}
5	6.0714 ^{BA}	6.1054 ^{BDAC}	6.0982 ^{BA}	6.0071 ^{BA}
6	6.1179 ^{BA}	6.1429 ^{BDAC}	5.8304 ^{BA}	6.1125 ^{BA}
7	6.3054 ^{BA}	6.3393 ^{BDAC}	6.1411 ^{BA}	6.3750 ^{BA}
8	6.3429 ^{BA}	5.5232 ^{DEC}	5.3446 ^{BC}	5.6429 ^{BC}
9	6.0161 ^{BA}	5.5929 ^{DEC}	5.6268 ^{BAC}	5.6911 ^{BC}
10	6.2714 ^{BA}	6.025 ^{BDAC}	5.9143 ^{BA}	6.0071 ^{BA}
11	6.2482 ^{BA}	6.0446 ^{BDAC}	6.1357 ^{BA}	6.1286 ^{BA}

12	6,7161 A	6.5518 BA	6.1143 BA	6.3464 BA
13	6.5589 BA	6.4286 BAC	6.1321 BA	6.2929 BA
14	6.5804 BA	6,6804 A	6,5375 A	6,7089 A
DMS	0,8322	0,9358	0,9911	0,9228

[1] Averages with the same letter in the same column did not differ (p<0, 05), n=56 tasters.

The results showed a significant difference ($p<0.05$) between the various attributes assessed.

For the AROMA attribute, only sample 12 (redistilled and with 2% added sugar) differed significantly from sample 1 (traditional, with no added sugar).

Although there were no significant differences between the other samples, the averages (Figure 10) show a trend towards greater acceptance of the Aroma attribute when comparing traditional and redistilled samples, since 4 of the 7 highest averages are redistilled samples. Still on the subject of aroma, it's worth noting that among the 7 highest averages, 5 samples have concentrations equal to or greater than 1.5% sugar, showing the influence of added sugar on the aroma attribute as well.

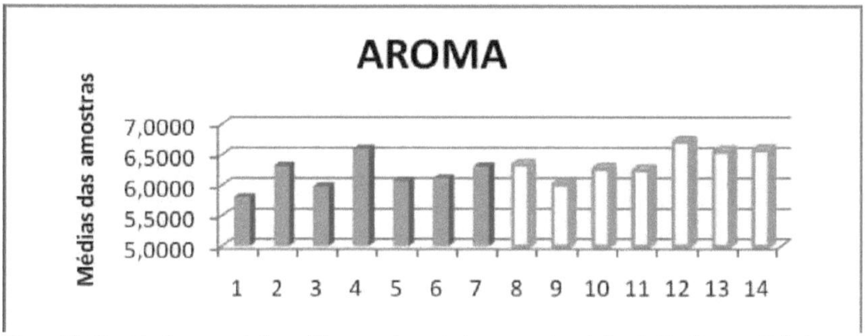

Figure 10: Graphical representation of the acceptance test averages in relation to the flavour attribute.

With regard to the results obtained for the flavour attribute (Figure 11), significant differences can be seen between samples 14, 12 and 13 (redistilled and with sugar concentrations between 2 and 3%) and samples 3 and 1 (traditional, without sugar and with 1% added).

Still in relation to the flavour attribute, it can be seen that of the 7 samples that received the highest averages, all had sugar concentrations equal to or greater than 1.5%, confirming that the addition of sugar has a major influence on the sensory evaluation of cachaça.

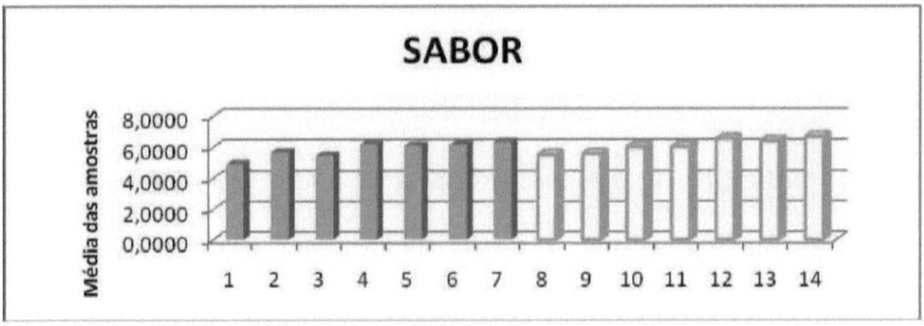

Figure 11: Graphical representation of the acceptance test averages in relation to the flavour attribute.

With regard to the sweetness attribute, the results (table 1) show a significant difference between samples 14 (redistilled and with 3% added sugar) and samples 3, 8 and 1, with samples 3 and 1 being traditional cachaças (without sugar and with 1% added) and sample 8 being redistilled and without added sugar.

Analysing the results shown in Figure 12, it can be seen that of the 7 highest averages, 5 are redistilled and 6 of them have sugar concentrations greater than or equal to 1.5%.

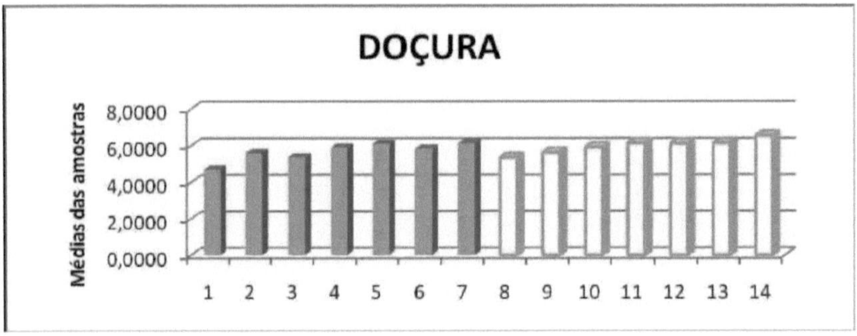

Figure 12: Graphical representation of the acceptance test averages in relation to the sweetness attribute.

Based on the results obtained for the overall impression attribute, there are significant differences between samples number 14 (redistilled and with 3% added sugar, the highest rated), samples 9 and 8 redistilled (without added sugar and with 0.6%) and samples 3 and 1 traditionally distilled (also without sugar and with 1%).

Still on the overall impression attribute (Figure 13), it can be seen that the 7 highest averages refer to samples with sugar concentrations greater than or equal to 1.5% and that 4 of them are redistilled samples.

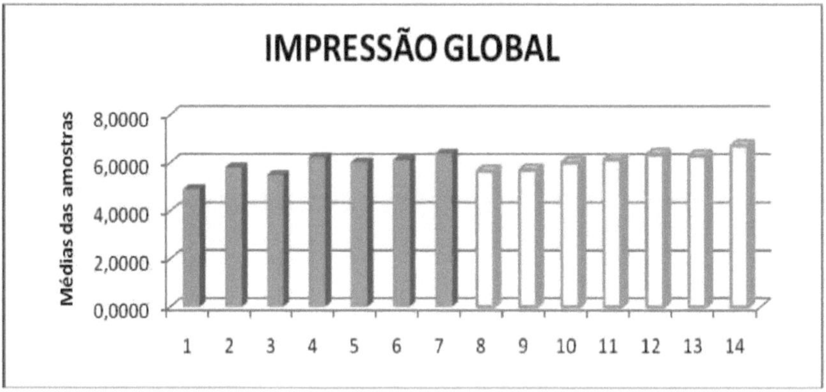

Figure 13: Graphical representation of the acceptance test averages in relation to the overall impression attribute.

Finally, analysing the results obtained in relation to purchasing attitude (Figure 14), we can see that samples 7 and 12 were more acceptable, as most of the judges would buy these cachaças. Samples 1 and 8, on the other hand, were the ones that obtained the worst concepts for buying attitude, being at the top of those that the judges would certainly not buy.

The samples with the worst purchase rates are precisely those without added sugar, proving once again the important role of sugar in the acceptance of this product.

It can also be seen that the redistilled samples were the ones that obtained the greatest acceptance and the highest level of purchase compared to the traditional samples, also showing the importance of this process in the sensory quality of cachaça.

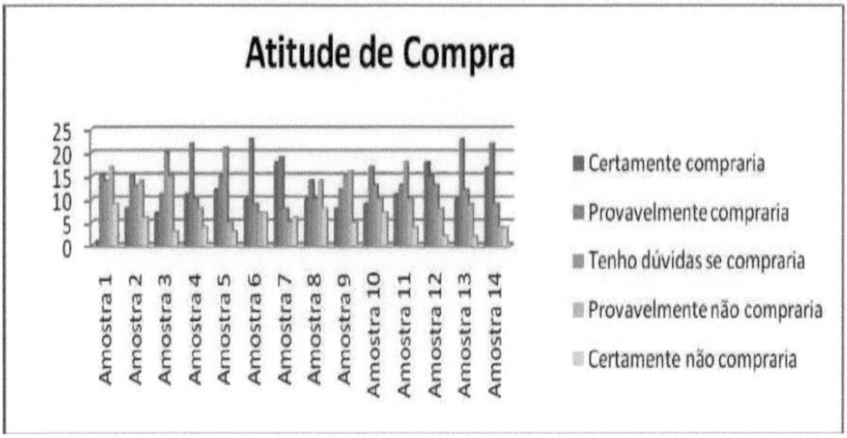

Figure 14: Graphical representation of the purchasing attitude of the samples analysed.

Table 2 shows the results of the levels of copper, higher alcohols, ethyl acetate, acetaldehyde, methanol, volatile acidity and real alcoholic strength present in the traditional and redistilled samples without added sugar.

Table 2 - Content of some compounds in the samples

Parameters	Sample 1 (traditional)	Sample 8 (redistilled)	Unit
Copper	0,50	0,60	mg/L
Higher alcohols	254,0	266,1	mg/100 mL p.a.
Ethyl Acetate	5,1	8,8	mg/100 mL p.a.
Acetaldehyde	3,4	5,7	mg/100 mL p.a.
Methanol	4,1	4,6	mg/100 mL p.a.
Volatile acidity	28,7	4,4	mg acetic acid/100 mL a.a.
Real alcoholic strength	36,55	38,01	°GL

The volatile acidity parameter (Figure 15) was the one that showed significant differences between the samples, with sample 1 (traditional) having the highest value (28.7 mg acetic acid/100 mL a.a.) and sample 8 (redistilled) the lowest (4.4 mg acetic acid/100 mL a.a.). In Figure 15, this reduction, which certainly has everything to do with the better acceptance of the redistilled cachaça samples, can be seen more clearly.

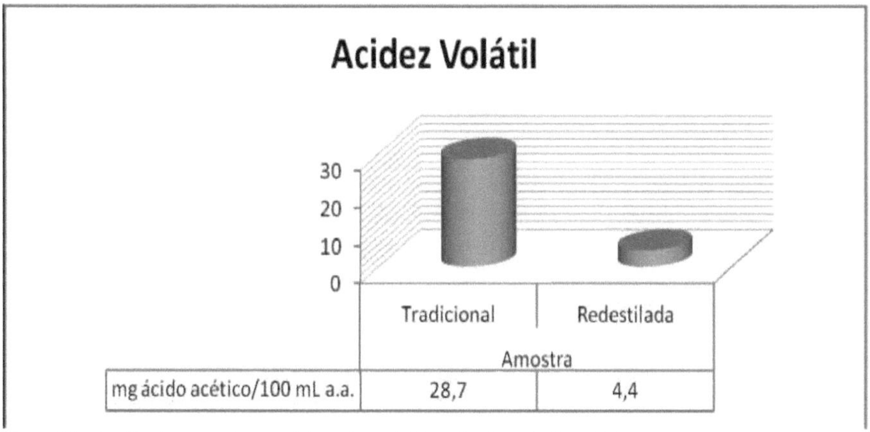

Figure 15: Graphical representation of the values obtained for the volatile acidity parameter.

CONCLUSIONS

The process of redistilling the cachaça also showed a significant sensory improvement, since the redistilled samples always showed higher averages than the corresponding samples obtained traditionally, indicating that this is a valid option for checking any defects in cachaça obtained under unsuitable conditions.

The results of the physico-chemical analyses corroborate the results obtained in the sensory analysis, justifying the better acceptance scores for the redistilled samples due to the reduction in volatile acidity, a factor that is known to be important for the sensory quality of cachaça.

The results obtained confirm that the practice of adding sugar to cachaça has a positive influence on consumer acceptance of the final product, both in relation to traditionally produced cachaças and also in the case of redistilled samples.

ACKNOWLEDGEMENTS

To Capes for the financial support.

MAPA (Ministry of Agriculture, Livestock and Supply) for the physical and chemical analyses.

BIBLIOGRAPHICAL REFERENCES

BRAZIL. Ministry of Agriculture. Ordinance no. 76 of 26 November 1986. Provides for the analytical methods of beverages and vinegar. **Official Gazette of the Federative Republic of Brazil**, Brasília, 28 Nov. 1986. Section 1, pt. 2.

BRAZIL, Normative Instruction No. 13 of 30 June 2005. Approves the Technical Regulation for Setting the Standards of Identity and Quality for Aguardente de Cana and Cachaça. Ministry of Agriculture, Livestock and Supply. **Federal Official Gazette,** DF, 30 June 2005, Section 1.

CARDELLO, H.M.A.B.; FARIA J.B.-"Quantitative descriptive analysis of sugar cane spirit during ageing in oak barrels *(Quercus alba* L.)" **- Ciência e Tecnologia de Alimentos -** v. 18 n. 2 Campinas May/Jul. 2000.

COMMISSION REGULATION (EC) No 2870/2000 of 19 December 2000 laying down Community reference methods for the analysis of spirit drinks THE COMMISSION OF THE EUROPEAN COMMUNITIES, **Official Journal of the European Union**.

MAÇATELLI, M. **"Determining the Sensory Profile of Commercial Brands of Cachaça"** 2007. 126p. Dissertation (Master's Degree in Food Science) - Universidade Estadual Paulista "Júlio de Mesquita Filho". Faculty of Pharmaceutical Sciences, Department of Food and Nutrition, Araraquara Campus. 2007.

ELECTRONIC REFERENCE

ALVES, M. | **Mercado de Cachaça** - Envasador Magazine - Issue 40 - Supplier Guide Mon, 02 March 2009 00:00 - Available at: http://www.ibrac.net. Accessed on: 15/09/2010.

CHAPTER 4

FINAL CONCLUSIONS

The results obtained in the sensory tests showed that the process of redistilling cachaça represented a significant sensory improvement, since the redistilled samples always had higher averages than the corresponding samples obtained traditionally, indicating that this is a valid option for improving the sensory quality of sugar cane spirits.

The results obtained in the physico-chemical analyses also explain, at least in part, those obtained in the sensory analysis, as seen in the better acceptance scores for the redistilled samples.

Among the parameters analysed, the obvious reduction in volatile acidity stands out, making it possible to relate this factor, which is known to be important for the sensory quality of cachaça, as being directly related to the positive effect of the redistillation process.

Likewise, the results obtained allow us to affirm that the addition of sugar to cachaça undoubtedly has a positive influence on consumer acceptance of the final product, reducing the perception of defects related to inadequate production techniques.

BIBLIOGRAPHICAL REFERENCES

AMERINE, M.A. Principles of sensory evaluation of food. **New York Academic**, 1965. p.349-397.

AQUINO, F. W. B, **Dextran in sugars and cane spirits**. Thesis (PhD) - University of São Paulo, São Carlos, 2009.

BIZELLI, L. C.; RIBEIRO, C. A. F.; NOVAES, F. V. Double distillation of sugar cane spirit: total acidity and copper content. **Scientia Agrícola**, Piracicaba, v. 57, n. 4, p. 623627, 2000.

BRAZIL, Normative Instruction No. 13 of 30 June 2005. Approves the Technical Regulation for Setting the Standards of Identity and Quality for Aguardente de Cana and Cachaça. Ministry of Agriculture, Livestock and Supply. **Federal Official Gazette,** DF, 30 June 2005, Section 1.

CANTÃO, F. O. **"Physical-chemical analyses and evaluation of the presence of copper in sugarcane spirits using aluminosilicates"** - Master's thesis, UFLA, 2006.

CARDELLO, H.M.A.B.; FARIA J.B.-"Quantitative descriptive analysis of sugar cane spirit during ageing in oak barrels (Quercus alba L.)" - **Ciência e Tecnologia de Alimentos** - v. 18 n. 2 Campinas May/July 1998

CARDOSO, D. R.; ANDRADE-SOBRINHO, L.G.; LIMA-NETO, B.S.; FRANCO, D.W. A rapid and sensitive method for dimethylsulphide analysis in brazilian sugar cane spirits and other distilled beverages. **Journal of the Brazilian Chemical Society,** v. 15, n. 2, p. 277-281, 2004

FARIA, J. B; LOYOLA, E.; LOPES, M.G. ; DUFOUR, J.P. . **Cachaça, Pisco and Tequila**. In: LEA, A G H; PIGGOTT J R. (Org.). Fermented Beverage Production 2nd. New York: Kluwer Academic / Plenium Publishers, 2003, v., p. 335-363.

FARIA, J. B.; FERREIRA, V.; LOPEZ, R.; CACHO, J. The sensory characteristic defect of "cachaça" distilled in absence of copper. **Revista Alimentos e Nutrição**, Araraquara, v. 14, n. 1, p. 1-7, 2003

FARIA, J. B. **Determination of the compounds responsible for the sensory defect of cane spirits distilled in the absence of copper**. 2000. 94 f. Thesis (Livre- Docência). Faculty of Pharmaceutical Sciences - UNESP - Araraquara- SP.

FARIA, J.B.; FRANCO, D.W.; CARDELLO, H.M.A.B.; BÔSCOLO M.; LIMA NETO, B.S. **Sensory evaluation of sugar cane brandy (Saccharum officinarum l.) during ageing in oak barrels (Quercus sp).** Ver. Brás. Anal. Alim., v.1, p.7 - 14, 1995.

FRANCO, A. C. **Redistillation of Cachaça** Dissertation presented to the Faculty of Pharmaceutical Sciences of Araraquara - UNESP - to obtain a Master's Degree in the area of Food Sciences and Nutrition. 2008.

FRANCO, A. C.; ROTA, M. B.; FARIA, J. B. Redistillation of Cachaça - **Alimentos e Nutrição** - v. 20, n.2, p. 331-334, 2009.

FORLIN, F.J. **Maturation of sugar cane spirit composed with oak wood extract in polyethylene terephthalate (PET) packaging.** Thesis (doctorate) - State University of Campinas. Faculty of Food Engineering, 2005.

ADOLFO LUTZ INSTITUTE. Analytical standards of the Adolfo Lutz Institute. 3.ed. São Paulo, SP, 1985. v.1, 533p.

ISIQUE, W.D.; LIMA NETO, B.S.; FRANCO, D.W. The presence of polyphenols in flocs from sugar cane spirit. **BRAZILIN MEETING ON THE CHEMISTRY OF FOOD AND BEVERAGES**, 2., 1999, Araraquara: F.C.F., 1999.p.2.

JANZANTTI, N. S. **"Volatile compounds and flavour quality of cachaça"**. Thesis (Doctorate) - State University of Campinas. Faculty of Food Engineering, 2004.

LIMA, U. A. **"Study of the main factors affecting the components of the non-alcohol coefficient of sugar cane spirits"**. 1964. 141p. Thesis (full professor). Luiz de Queiroz College of Agriculture, Piracicaba: University of São Paulo.

LIMA, U.A. Brandy. In: AQUARONE, E.; LIMA, U.A.; BORZANI, W. **Alimentos e bebidas**

produzidos por fermentação (Série biotecnologia). São Paulo: Edgard Blucher, 1983. v.5, chap.4, p. 79 - 102.

LIMA, U. A. **Manufacture in small distilleries**. Piracicaba: Fundação Estudos Agrários "Luiz de Queiroz", 1999. 187 p.

MAÇATELLI, M. **"Determining the Sensory Profile of Commercial Brands of Cachaça"** 2007. 126p. Dissertation (Master's Degree in Food Science) - Universidade Estadual Paulista "Júlio de Mesquita Filho". Faculty of Pharmaceutical Sciences, Department of Food and Nutrition, Araraquara Campus. 2007.

MIRANDA, M. B; MARTINS, N. G. S; BELLUCO, J. H; ALCARDE, A. R. "Chemical quality of Brazilian cachaças and brandies" **Ciências Tecnologia de Alimentos**, Campinas, Oct.-Dec. 2007

NÓBREGA, I.C.C. - "Analysis of the volatile compounds in sugar cane spirit by dynamic headspace concentration and gas chromatography-mass spectrometry" **Ciências e Tecnologia de alimentos**, Campinas, 23(2), p.210 - 216, May-August 2003.

NOGUEIRA, A.M.P.; FILHO, W.G.V. - **"AGUARDENTE DE CANA"** - UNESP Campus De Botucatu Faculdade De Ciências Agronômicas, 2005.

NOVAES, F.V. **"Production and quality of sugar cane brandy"**. Piracicaba: ESALQ, 1995. 27p. Handout.

NOVAES, F. V. **Basic notions on distillation theory.** Piracicaba: ESALQ/Depto de Ciência e Tecnologia Agroindustrial, 1994. 22p

PIGGOTT, J. R. Ç.; SHARP, R. Ç.; DUNCAN, R. E. B. **The Science and Technology of Whiskies**. New York: Longman Scientific & Technical, 1989. 410 p.

ORTEGA, C.;LOPEZ, R.;CACHO, J.; FERREIRA, V. **"Fast analysis of important wina volatile compounds: development and validation of a new method based on gas chromatographic-flame ionisation detection analysis of dichloromethane micro extracts"**. J. Chromatogr. A, v. 923, p. 205 - 214, 2001.

COMMISSION REGULATION (EC) No 2870/2000 of 19 December 2000 laying down Community reference methods for the analysis of spirit drinks THE COMMISSION OF THE EUROPEAN COMMUNITIES, **Official Journal of the European Union**.

ROTA, M. B. **Effect of the bidistillation process on the sensory quality of cachaça.** 2008. 80p. Dissertation (Master's in Food Science) - Universidade Estadual Paulista "Júlio de Mesquita Filho". Faculty of Pharmaceutical Sciences, Department of Food and Nutrition, Araraquara Campus. 2008.

SAS Institute. **SAS user's guide:** statistics. Cary, 1993.

SIELBERG, K.J. **Effects of protein - polyphenol interactions on berevage haze**. Food Chem., v.47, p.353-362, 1999.

YOKAYA, F. **"Fabricação da aguardente de cana"** (Industrial fermentations series, n°2). Campinas, SP: Fundação Tropical de Pesquisa e Tecnologia "André Tosello", 1995. 93 p.

ELECTRONIC REFERENCE

ALVES, M. | **Mercado de Cachaça** - Envasador Magazine - Issue 40 - Supplier Guide Mon, 02 March 2009 00:00 - Available at: http://www.ibrac.net. Accessed on: 15/09/2010.